Cover art and promotional copy (c) Booklife 2011

ISBN 978-1-105-05337-5

TG No. 36   October 2009

TECHNICAL GUIDE NO. 36

PERSONAL PROTECTIVE MEASURES AGAINST INSECTS AND OTHER ARTHROPODS OF MILITARY SIGNIFICANCE

TABLE OF CONTENTS

APPENDICES IV
LIST OF TABLES IV
LIST OF FIGURES V
1. INTRODUCTION 1
   1-1. Purpose 1
   1-2. References 1
   1-3. Suggested Improvements 1
   1-4. Background 2
   1-5. Arthropods of Military Significance 3
   1-6. Responsibilities 4

2. METHODS OF PROTECTION 8

SECTION I. INTRODUCTION 8
   2-1. General 8

SECTION II. AVOIDANCE 8
   2-2. Field Strategies 8
   2-3. Information Sources 8
   2-4. Emergency Requisition of Repellents and Pesticides 9

SECTION III. PHYSICAL BARRIERS 9
   2-5. Clothing 9
      a. Field Uniform 9
      b. Tick Checks 11
      c. Chiggers 13

2-6. Protective Equipment...13

    a. Introduction. 13

    b. Insect Head Net. 14

    c. Insect Protective Mesh Parka and Mittens 16

    d. Insect Net Protectors 17

SECTION IV. REPELLENTS.. 19

2-7. Introduction 19

2-8. Deet (N,N-diethyl-m-toluamide or N,N-diethyl-3-methylbenzamide). 20

    a. Introduction.. 20

    b. Health and Safety Considerations 20

    c. Formulations 21

        (1) Two-Ounce Tube 21

        (2) Camouflage Face Paint..23

        (3) Insect Repellent With Sunscreen 24

        (4) Insect Repellent Stick.. 24

2-9. Permethrin 26

    a. Introduction.. 26

    b. Health and Safety Considerations 28

    c. Methods of permethrin treatment... 29

        (1) Individual Dynamic Absorption (IDA) Kit.. 29

        (2) Aerosol Spray Can 36

        (3) 5.1-Ounce (151 ml) Bottle 38

        (4) Factory Treatment of military uniforms... 42

2-10. Miscellaneous Repellent. 43

2-11. DoD Insect Repellent System.. 45

2-12. Area Repellents.. 45

2-13. Repellent Devices that are Worn on the Body 45

    2-14. Introduction 47
    2-15. Commercial Products 47
        a. Non-deet Products 47
        b. Ingested Products 47

SECTION VI. PESTICIDE REDUCTION THROUGH PHYSICAL/MECHANICAL MODIFICATIONS AND SANITATION 48
    2-16. DoD and IPM 48

SECTION VII. CONCLUSION 48
    2-17. Summary 48
    2-18. Training Package 49

APPENDICES

APPENDIX A - REFERENCES 50

LIST OF TABLES

Table 1. Arthropods of Military Importance and the Major Diseases They Transmit 5

LIST OF FIGURES

Figure 2-1. Proper wearing of field uniform minimizes exposure to arthropod attack 10

Figure 2-2a. Buddy-System check for ticks 12

Figure 2-2b. Buddy-system tick removal with tweezers 12

Figure 2-3. Remove ticks using a lint roller 12

Figure 2-4. Remove ticks using a ring of tape 13

Figure 2-5. Insect Head Net NSN 8415-00-935-3130 photographed 14

Figure 2-6. Insect Head Net, NSN 8415-00-935-3130 photographed over helmet 15

Figure 2-7. Remove insect net from the back 15

Figure 2-8. Insect head net being treated with permethrin aerosol can 16

Figure 2-9. Insect Protective Mesh Parka 16

Figure 2-10. Insect net protector (Mosquito Bed Net), NSN 7210-00-266-9736 on folding cot 18

Figure 2-11. Standard Military Skin Repellent, NSN 6840-01-284-3982 22

Figure 2-12. Camouflage Face Paint 23

Figure 2-13. Insect Repellent With Sunscreen 24

Figure 2-14. Insect Repellent Stick, NSN 6840-00-142-8965, ... 25

Figure 2-15. Permethrin IDA Kit, NSN 6840-01-345-0237 30

Figure 2-16. IDA kit instructions for treating coat 31

Figure 2-17. IDA kit instructions for treating trouser 32

Figure 2-18a-d. Steps 1-4 in using the IDA Kit, NSN 6840-01-345-0237 33

Figure 2-19a-d. Steps 5-8 in using the IDA Kit NSN 6840-01-345-0237 34

Figure 2-19e. Step 9 in using the IDA Kit NSN 6840-01-345-0237 35

Figure 2-20. Permethrin aerosol, NSN 6840-01-278-1336 37

Figures 2-21a and 2-21b. Applying permethrin aerosol to the field uniform 38

Figure 2-22. Permethrin 5.1-ounce (151-ml) bottle, NSN 6840-01-334-2666 39

Figure 2-23. Applying permetrin by air-compressed Sprayer to multiple field uniforms 41

Figure 2-24. Applying permetrhin by air-compressed sprayer to insect net protector 42

Figure 2-25. Applying permetrhin by air-compressed sprayer to external surface of a tent 42

Figure 2-26. CHIGG-AWAY® 188-ml plastic squeeze bottle, NSN 6840-01-137-8456, ... 45

Figure 2-27. Skin lesions on the legs caused by human use of flea and tick collars 47

TG No. 36   October 2009

TECHNICAL GUIDE NO. 36

PERSONAL PROTECTIVE MEASURES AGAINST
INSECTS AND OTHER ARTHROPODS OF MILITARY SIGNIFICANCE

1.1 INTRODUCTION

1-1. Purpose

This Technical Guide (TG) provides preventive medicine (PVNTMED) information and guidance to Department of Defense (DoD) personnel who may come into contact with nuisance or disease-carrying arthropods (disease vectors), or who are responsible for protecting the health of personnel. It describes the DoD Insect Repellent System and other techniques that provide maximum, safe protection from arthropod attack. These techniques include the use of protective clothing and equipment, repellents, pesticides, and other strategies. This AFPMB TG is based on, and supersedes, U.S. Army Environmental Hygiene Agency (USAEHA) TG No. 174, Personal Protective Techniques Against Insects and Other Arthropods of Military Significance, June 1991, and the Armed Forces Pest Management Board TG No. 36, Personal Protective Techniques Against Insects and Other Arthropods of Military Significance, May 2006 version.

1-2. References

References are listed in Appendix A.

1-3. Suggested Improvements

The organization responsible for this TG is the Armed Forces Pest Management Board (AFPMB). Users are invited to send comments and suggested improvements either by e-mail (www.afpmb.org) or mail directly to:

Information Services Division
AFPMB
WRAMC, Forest Glen Annex, Bldg. 172
6900 Georgia Avenue, NW
Washington, DC 20307-5001
(301) 295-7476 or DSN 295-7476

or by accessing the following link: http://www.afpmb.org/forums/sendmessage.php

1-4. Background

  a. Historically, in every war and military conflict, combat power has been reduced more significantly by disease and non-battle injuries than from direct combat casualties. A large number of diseases affecting the troop strength of deployed units is directly attributed to disease-carrying arthropods. Moreover, arthropods can inflict severe physical, psychological, and economic stresses that threaten the military mission. Not only do they transmit disease, but the bites they inflict can be painfully distracting and can lead to devastating secondary infections, dermatitis, or allergic reactions. Further, contamination of food and damage to other commodities are costly.

  b. History is replete with examples of how arthropod-borne diseases have significantly impacted military operations (Bunn et al., 1955; Dickens, 1990; Gambel ,1995; Plorde, 1983).

   (1) In 1812, Napoleon invaded Russia with 422,000 men. Within three months, seven of every ten soldiers had fallen to epidemic louse-borne typhus, leaving a force of only 100,000. Cold injuries completed the devastation of the disease-weakened force, and by the time Napoleon's *Grande Armée* retreated out of Russia only 10,000 remained. Dysentary and pneumonia joined with typhus to further reduce the force to fewer than 3,000 living troops.

   (2) General George Washington's Continental Army experienced ten soldiers dead of diseases for every combat casualty. In the War of 1812, General Andrew Jackson's victory at the Battle of New Orleans in January 1815 was immediately followed by the death of most of the surviving American and British troops from mosquito-borne malaria. During the Civil War, there was a 2:1 ratio in deaths from disease versus combat. In 1898 during the Spanish-American war, Army disease casualties included 90,416 cases of malaria, 1,169 cases of yellow fever and 249 cases of dengue fever, with respective fatality rates of 4, 123, and 8 per 1,000 cases.

   (3) During World War I, the ratio of deaths from disease versus combat in U.S. troops improved to 1:1, but there were still 16,930 cases of malaria. During World War II, it is estimated that over 24,000,000 man-days were lost to arthropod-borne diseases. An entire regiment was rendered ineffective by scrub typhus. Dengue fever reached a high of 28,292 cases in the Southwest Pacific with 52 cases per 1,000 troops per year. An outbreak of dengue in the New Hebrides in 1943 affected 26 percent of U.S. forces (more than 6,000 personnel). During the Korean War, U.S. Army forces suffered more than 30,000 cases of malaria, and hundreds of Americans were hospitalized during a 1951 epidemic of Korean hemorrhagic fever. During the Vietnam War, diseases accounted for 56 to 74 percent of all U.S. Army hospital admissions. From 1965 through 1970, the Army lost 2,000,000 man-days to malaria alone. Units operating in the Ia Drang Valley in 1965 reported an annual malaria rate of 600 cases per 1,000 troops, which rendered two entire battalions ineffective for a time. Annual epidemics of Japanese encephalitis were also devastating.

   (4) In 1993, over 200 cases of malaria were reported among U.S. military personnel who served in Somalia during Operation Restore Hope. Dengue virus infections occurred in military personnel stationed in Haiti as part of Operation Uphold Democracy in

1994, and cases of leishmaniasis were an outcome of military operations in Central and South America, and the Middle East in the 1990s.

(5) During Operation Iraqi Freedom, there were 653 cases of cutaneous and 2 cases of confirmed visceral leishmaniasis by the close of March 2004. Some estimates have placed the actual number much higher; figures ranging up to 2,500 have been cited which would be an infection of 1% of US service members serving in Iraq during 2003 – 2004 (Korzeniewski and Olszanski, 2004).

(6) In the United States, tick-borne infections such as Lyme disease continue to have a significant impact on morbidity of service members training in areas of the Southeast, Northeast, and upper Midwest; in addition, newly emerging infections such as the human ehrlichioses are now posing further hazards. Nuisance arthropod bites and the diseases they transmit will continue to be a serious threat to service members in training and in combat.

1-5. Arthropods of Military Significance

a. Table 1 lists the major arthropod pests of military importance and the primary diseases that they transmit (Heymann, 2004; Speilman and James, 1990). This section is only intended to be a simple introduction to the topic. The AFPMB website (www.afpmb.org) contains an electronic library of publications pertaining to medical entomology from general entomology information to specific journal articles on a myriad of vector species. Information can be retrieved by accessing the website and then clicking on the icon "Search Literature Database" and following the instructions or by simply accessing the following link:

http://lrs.afpmb.org/rlgn_app/ar_login/guest/guest

In addition, the U.S Army Center for Health Promotion and Preventive Medicine has a websiJ te devoted to military entomological issues which can be accessed at the following link:

http://usachppm.apgea.army.mil/ento/

Readers are also encouraged to review the following references for detailed information on arthropod vectors: Mullen and Durden (2009), Goddard (2007) Elridge and Edman (2004), Kettle (1995), or Rossignol and Feinsod (1990).

(1) In most regions of the world, mosquitoes are the foremost disease vectors and nuisance pests. They transmit three of the most serious vector-borne diseases that jeopardize U.S. forces: malaria, dengue, and viral encephalitis.

(2) Phlebotomine sand flies transmit other major diseases of military importance such as sand fly fever and leishmaniasis.

(3) Other arthropods that cause disease, nuisance problems, or direct injury, are black flies, deer flies, horse flies, stable flies, tsetses, horse and deer flies, filth flies, bot flies, Tumbu

(a) Ticks can host a broad range of pathogens, including the agents of Lyme disease, Rocky Mountain spotted fever, and the human ehrlichioses.

(b) Although some arthropods, notably filth flies, do not bite and are therefore not true biological vectors of disease, they can mechanically transmit many serious illnesses such as d ysentery, cholera, salmonella, shigellosis, and typhoid fever. Additionally, they are often numerous enough in many areas to pose an extreme nuisance, constantly seeking moisture from sweat and from fluids of the eyes, nose, and mouth.

(c) Other arthropods that directly cause human injury, but are also not true vectors of disease, are the bot flies and the Tumbu fly. Larvae (also known as maggots) of these flies burrow into human skin and develop in the tissue, causing intense pain and itching. Invasion of tissue by fly maggots is called myiasis.

1-6. Responsibilities

a. Personal protection is an individual responsibility, although it is also an important adjunct to unit-level and higher echelon preventive medicine countermeasures. Military personnel must be aware of the following:

(1) Types of arthropods in an area;

(2) Their habits;

(3) The threat they present;

(4) The resources available for protection;

(5) How to use these resources effectively.

b. Command emphasis is essential! Each of the three services of the DoD provide guidance and policy formulation that ultimately tasks Commanders and medical personnel with ensuring that personal protective strategies are in place, that all appropriate protective resources are being provided, and that individuals are using these protective resources properly (US Army Regulation 40-5; US Navy P-5010-8, US Air Force Instruction 48-102). U.S. Army Pamphlet 40-11 (DA PAM 40-11) 2005 requires all company and battery-sized units to pre-stock specific quantiJ ties of the standard military skin and clothing repellents for each individual. It is imperative that service members have immediate access to sufficient personal protective supplies if they are to

Table 1. Arthropods of Military Importance and the Major Diseases They Transmit

Visual ID	Common Name	Genus	Diseases
	Biting midges	*Culicoides*	- Visceral filariasis (mansonellosis) - Oropouche fever
	Body lice	*Pediculus*	- Epidemic typhus - Relapsing fever - Trench fever
	Black flies	*Simulium*	- Onchocerciasis (river blindness)
	Bot flies	*Dermatobia*	- Myiasis
	Deer flies	*Chrysops*	- Eye worm disease (loa loa) - Tularemia

TG No. 36　　　　　　　　　　　　　　　　　　　　　　　　　　　October 2009

Table 1. Arthropods of Military Importance and the Major Diseases They Transmit (Continued)

Visual ID	Common Name	Genus	Diseases
	Fleas1	*Xenopsylla*	- Plague - Murine typhus
	Kissing bugs	*Rhodnius,* *Triatoma,* *Panstrongylus*	- Chagas' disease (American trypanosomiasis)
	Mites		
	Chigger mites	*Leptothrombidium*	- Scrub typhus
		Sarcoptes	- Scabies
	Mouse mites	*Lyponyssoides*	- Rickettsialpox
	Mosquitoes	*Aedes*	- Dengue - Yellow fever - Viral encephalitis
		Anopheles	- Malaria
		Culex, Aedes	- Viral fever (Oropouche, Rift Valley, Chickungunya) - Lymphatic filariasis (Wuchereriasis, Brugiasis)

6

Table 1. Arthropods of Military Importance and the Major Diseases They Transmit (Continued)

Visual ID	Common Name	Genus	Diseases
	Sand flies	*Lutzomyia,* *Phlebotomus*	- Leishmaniasis - Sand fly fever - Bartonellosis
	Ticks		
	Hard ticks	*Dermacentor*	- Spotted fevers - Colorado tick fever
		Ixodes	- Lyme disease - Babesiosis - Viral encephalitis - Tularemia
		Amblyomma *Ixodes*	- Human ehrlichioses
		Hyalomma	- Crimean-Congo hemorrhagic fever
	Soft ticks	*Ornithodorus*	- Relapsing fever
	Tsetse flies	*Glossina*	- Trypanosomiasis (African sleeping sickness)
	Tumbu flies	*Cordylobia*	- Myiasis

2.1 METHODS OF PROTECTION

Section I. Introduction

2-1. General

Arthropod-borne diseases and nuisance pests can be prevented or controlled by using a number of techniques including personal protective measures and environmental controls. In many situations, personal protective measures such as avoiding infested areas, or the use of physical barriers or chemical repellents, may be the only means of protection available. Environmental controls, while not a primary focus of this TG, are nevertheless mentioned to illustrate the total integrated approach that should be employed by a unit in field situations. They include such measures as sanitation, mechanical and behavioral modifications, and pesticide application.

Section II. Avoidance

2-2. Field Strategies

The most effective and obvious means of preventing exposure to arthropods is to avoid their known habitats. Absolute avoidance of arthropod pests is often neither practical nor possible. If the tactical situation allows, choose bivouac sites that are dry, open, and as uncluttered as possible. Avoid sites with rodent burrows and proximity to local settlements, animal pens, and other areas where arthropod infestations are likely to be concentrated.

2-3. Information Sources

a. Preventive Medicine personnel should provide guidance on the presence of arthropod populations in an area based on information obtained through surveillance or via intelligence sources.

(1) The Information Services Division of the Armed Forces Pest Management Board (AFPMB) compiles Disease Vector Ecology Profiles (DVEPs), which are concise, comprehensive summaries of the vector-borne diseases that occur in specific countries or other geographic areas. The DVEPs focus on causative agents, vector importance, bionomics, behavior, and pesticide resistance, as well as provide basic information on the geography and customs of each country. They may be obtained from the Information Services Division, AFPMB, WRAMC. Forest Glen Annex, Bldg 172, 6900 Georgia Avenue, NW, Washington, DC 20307-5001, DSN 295-7476, commercial 301-295-7476, or from the AFPMB web site at http://www.afpmb.org or by accessing the following link:

http://www.afpmb.org/pubs/dveps/dveps.htm

(2) Up-to-date worldwide information on diseases and vectors may be obtained from the National Center for Medical Intelligence (NCMI), Fort Detrick, Frederick, MD 21702-5004, 301-619-7574, DSN 343-7574, or by accessing the following link:

TG No. 36   October 2009

(3) The Centers for Disease Control and Prevention provides current information on all communicable diseases including arthropod-borne diseases. Information can be accessed at the following link:

http://www.cdc.gov/

2-4. Emergency Requisition of Repellents and Pesticides

a. Deploying and/or deployed forces often need pesticides and pest management equipment on short notice. The Defense Logistics Agency has established an agency wide Customer Interaction Center (CIC) to help meet these needs.

b. For emergency procurement of pesticides, including repellents and pest management equipment, pesticide application equipment and pesticide protection equipment, etc.: Contact the Defense Logistics Agency CIC at 1-877-DLA-CALL [1-877-352-2255], DSN 661-7766 and verbally speak the phrase "Supply & Transportation" when so prompted for assistance. For credit card orders, verbally speak the phrase "Credit Card Purchases" when so prompted for assistance.

c. For technical inquiries or assistance regarding emergency procurement of pesticides, repellents, pest management material/equipment, and all chemicals, contact the Chemical Office, Defense Supply Center Richmond, at DSN 695-3995 or commercial (804) 279-3995 during normal duty hours [0800-1630 hrs eastern standard time (EST)] or via Government cell phone (804) 651-4630, anytime, 7 days a week. FAX numbers are either (804) 279-3653 or 804- 2 79-3971.

Section III. Physical Barriers

2-5. Clothing

a. Field Uniform

Clothing is the first direct line of personal defense against arthropods. Proper wearing of the field uniform is essential to minimize skin exposure (Figure 2-1). If the risk of heat stress is a factor in a particular environment, common sense or advice from medical/Preventive Medicine personnel should dictate when the following recommendations are not practical.

(1) Tuck the pant leg into the boot or into the sock. This forces non-flying pests such as ticks, stinging ants, and spiders to climb up the outside of the pant leg, thus decreasing access to the skin and increasing the likelihood of being seen.

(2) Wear the uniform with the sleeves down, wrist openings secured, and collar closed to help protect the arms and neck from attack. This is especially important from dusk until dawn when many mosquito species and other nocturnal blood feeders are active.

(3) It is difficult for attacking pests to bite through the uniform fabric unless it is pulled tightly against the skin. Therefore, the uniform should be worn loosely, with an undershirt worn

The undershirt should be tucked into the pants to decrease entry access by crawling arthropods at the waist line.

Figure 2-1. Proper wearing of field uniform minimizes exposure to arthropod attack

(4) The field cap and its brim help protect the head and face. Some biting insects tend to avoid the shaded area of the face under the cap's brim (Schreck, 1989). In areas heavily infested with flying pests, a head net can be used over the cap or helmet as shown in Figure 2-1.

b. Tick Checks

(1) When in tick-infested habitats, check clothing routinely, and use the buddy system to check areas of the body that cannot easily be seen during self-examination (Fig 2-2a).

(a) Ticks can be removed from clothing by hand. However, avoid crushing any Ticks with your fingernails because their body fluids may contain pathogens and therefore be infective. After removal, disposal may pose a problem. If returned to the immediate area, ticks may reattach to the clothing or attack another individual. They can be destroyed by placing them in alcohol or by securing them within a piece of folded tape.

(b) An adhesive lint roller (available from most post/base exchanges and commissaries) is a very efficient means of quickly removing large numbers of ticks from the uniform, especially the very tiny larvae, which may be present in clusters of several hundred (Figure 2-3).

(c) Ordinary masking tape, cellophane tape, or similar substitute, are useful to remove ticks from clothing. A ring of tape can be made around the hand by leaving the sticky side out and attaching the two ends. Ticks will adhere to the tape when it is dabbed against the clothing (Figure 2-4). The tape can then be folded carefully over the ticks to prevent their escape and discarded with the trash.

(2) Once clothing is removed, it is important to carefully check all areas of the body for evidence of ticks. Re-examine the clothing, inside and out, and remove and dispose of all ticks. If a tick is found attached to the body, seek medical attention for removal. If proper medical attention is not readily available, follow the guidelines posted in the following link to remove the tick: http://chppm-www.apgea.army.mil/ento/TickEduc/TickRemoval-April2006.pdf

Tweezers are the best means of removing attached ticks (Fig 2-2b). See NSN item link:

http://www.afpmb.org/pubs/standardlists/equipment/pdfs/3740-01-474-7377.pdf

To help combat the threat of tick-borne diseases to DoD personnel, the Entomological Sciences Program (ESP) of the U.S. Army Center for Health Promotion and Preventive Medicine (USACHPPM) provides a tick identification and testing service for DoD health clinics within the continental United States. The ESP analyzes ticks for evidence of infection with the agents of several tick-borne diseases. Best results are obtained from live ticks so do not kill the tick after removal from body. Details of the service can be viewed at the following link:

USACHPPM ESP can be contacted directly for information on how to access the program if in an overseas location:

http://chppm-www.apgea.army.mil/contactus/Wemail.asp

United States Army Center for Health Promotion & Preventive Medicine
5158 Blackhawk Road
Aberdeen Proving Ground, MD 21010-5403
(800) 222-9698
DSN: Dialing from within CONUS   584-4375
Dialing from OCONUS   (312) 584-4375

Figure 2-2a. Buddy-System check for ticks

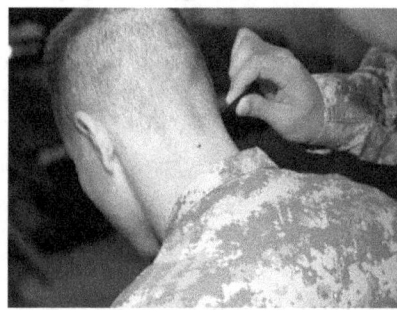

Figure 2-2b. Buddy-system tick removal with tweezers

Figure 2-3. Remove ticks using a lint roller

Figure 2-4. Remove ticks using a ring of tape

c. Chiggers

(1) The larval chigger mite is the stage that feeds on humans. These larvae are microscopic, and as such, are not seen without proper slide mounting and a microscope. The presence of chigger mites is generally not detected until the appearance of intensely itching bites.

(2) Wearing repellent-impregnated uniforms greatly reduces the likelihood of being infested with chiggers (Breeden et al., 1982). See Section IV.

(3) Bathing after field work, or as soon as operationally permissible, may reduce the severity of the resultant chigger infestation.

(4) Medical personnel may prescribe an anti-pruritic or antibiotic to help reduce itching or secondary infection.

d. Spiders, Scorpions, and Snakes

To reduce the chance of being bitten by spiders, scorpions (stung), and snakes:

(1) Always wear shoes or boots with the added protection of socks during waking hours.

(2) Never walk outdoors in bare feet, stocking feet or in flip flops.

(3) Shake out boots before putting them on.

(4) Store boots with socks pulled down over the boot tops as far as possible to prevent entry of such pests.

(5) Do not reach into concealed areas that might harbor spiders, scorpions, or snakes without carefully checking first.

2-6. Protective Equipment

a. Introduction.

Equipment items available through the military supply system can be used to augment the physical protection afforded by the physical clothing barrier. A current list of the available personal protection materiel can be accessed on the AFPMB website (www.afpmb.org) and then proceeding to "DoD Standard Pesticides and Pest Control Equipment Lists" then to "DoD Pest Management Materiel Other Than Pesticides" and then finally to "Section 19 - Personal Protection Materiel." The link to where one can access the DoD Standard Pesticide and Pest

Also, the AFPMB TG No. 24 (accessed at the AFPMB webiste) is the Contingency Pest Management Guide and will provide all the items on the contingency pest management list with stock numbers.

b. Insect Head Net.

(1) In areas heavily infested with flying pests, a head net can be used over the cap or helmet. The insect head net (NSN 8415-00-935-3130 or NSN 8415-00-935-2914) is a finely woven (30-mesh/inch), olive drab, nylon head covering that can be worn over the bare head, cap, helmet, or helmet liner (Figure 2-5). The cloth top piece has an elastic headband on the inside that fits securely over the head gear. A fabric-covered metal hoop holds the net away from the head and neck.

Figure 2-5. Insect Head Net NSN 8415-00-935-3130 photographed

(2) Put on the head net so that the elastic headband rests comfortably on the upper part of the forehead or grips close above the brim of the helmet (Figure 2-6). Tie the drawstring permanently so the drawstring knot is about 8 inches below the chin and the net fits snugly below the collar, both front and back. If breast pocket buttons are equipped on the subject uniform, hook the elastic loops found at the drawstring edge of the net over these buttons.

Figure 2-6. Insect Head Net, NSN 8415-00-935-3130 photographed over helmet

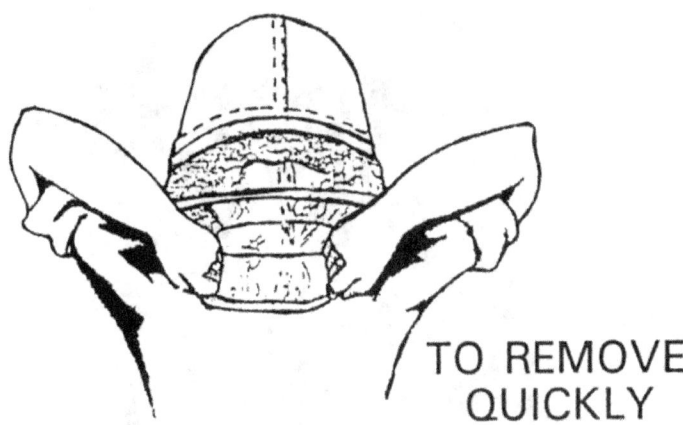

Figure 2-7. Remove insect net from the back

(3) For quick removal of the head net, grasp the back edge where it rests over the collar and pull forward over the head (Figure 2-7).

(4) The head net is especially useful in areas of very dense mosquito or biting fly populations. It may be worn while sleeping.

(5) For added protection, the head net may be lightly sprayed with permethrin [see Figure 2-8 and paragraph 2-9c(2)]. Allow it to dry thoroughly before wearing. The treatment should be effective for several months. In the absence of permethrin, the net may be hand-treated with deet repellent [see paragraph 2-8c(4)] every evening by dispensing a small quantity onto the palm of one hand (3 to 4 drops of the liquid, or a small dab of the lotion), rubbing the hands together to spread the repellent, and finally rubbing the netting between the hands. Repeat the process until all the netting has been lightly and evenly covered. It is not necessary to saturate the netting. **KEEP DEET REPELLENT OFF OF THE ELASTIC AS IT IS A PLASTICIZER AND MAY DAMAGE PLASTICS, RUBBER, VINYL, OR ELASTIC ITEMS.**

(6) Because of its small mesh size, the insect head net can be very hot for the wearer or may obscure vision, making it impractical in some climates and under certain deployed

Figure 2-8. Insect head net being treated with permethrin aerosol can

c. Insect Protective Mesh Parka and Mittens.

(1) A mesh parka or over jacket (Figure 2-9) is available (small, medium, large, extra large, extra-extra large: NSN 8415-01-483-2988; -3002; -3004, -3007, -3008 respectively) thJ at is effective <u>without</u> applying repellent to it, unlike the old repellent parkas. It is made from narrow-mesh polyester netting, is worn over outer clothing, and is snag resistant. The small mesh size not only protects against mosquitoes bites, but also prevents bites from very small fl ying insects such as "no-see-ums", sand flies, black flies, and gnats. The parka is waist-length, has a pocket, long sleeves, a drawstring and a mesh hood that covers the face and head. Remember, this parka offers protection without being treated with repellent.

Figure 2-9. Insect Protective Mesh Parka

(2) Insect Net Mittens (NSN 8415-01-192-2357) are also available to augment protection from biting arthropods.

d. Insect Net Protectors

Indoor protection can be greatly enhanced by using bed nets, pop-up nets, and tent screens. Unlike head nets, parkas, and mittens, the mesh size of bed netting and tent screens is not fine enough to keep out all biting arthropods, especially biting midges and sand flies. Treating bed nets and tent screens with repellents can significantly reduce the ability of these arthropods to gain entry (McCain and Leach, 2007; Jinjiang et al., 1988; World Health Organization, 1989).

(1) There are currently two available bed nets (NOT including the "pop-up" bed nets discussed in section 2 below) on the NSN stock system. The insect net protector (Figure 2-10) (NSN 7210-00-266-9736) is a finely woven (27-mesh/inch), olive drab, nylon canopy that can be used with the folding cot, hammock, steel bed, or shelter half-tent.

http://www.afpmb.org/pubs/standardlists/equipment/pdfs/7210-00-266-9736.pdf

Another nylon insect protector (NSN 7210-00-266-9740) is also available; this item has slightly smaller dimensions yet can also be used with the folding cot.

http://www.afpmb.org/pubs/standardlists/equipment/pdfs/7210-00-266-9740.pdf

(a) The insect net protector should be erected and supported in such a way as to prevent contact of the net with the sleeping person. This will decrease the risk of mosquitoes and other blood-feeding flies from biting the individual through the net. NSN items 7210-00-267-5641 [Pole, Folding Cot, Insect Net Protector, unit of issue SE (set of four wood poles)], or NSN 7210-00-300-6950 [Rod Insect Net Protector, unit of issue EA (a T-shaped metal rod) can be used to erect a bed net. If using Rods, two are required to suspend a bed net. Rods can be pushed in the ground. Rods do not fit or directly attach to military cots. Two Clamps, Insect Net Protector Rod, 7210-00-359-4850 (unit of issue EA) are required to properly erect the insect net protectors if attaching to a folding cot.

(b) Do not leave net in contact with the ground as crawling arthropods may use it to gain access to the sleeping area. Tuck the net under the mattress or sleeping bag. Bed nets should be installed before dusk, the time when many mosquitoes become active.

(c) Prior to retiring, a check should be conducted for the presence of flying insects, such as mosquitoes, trapped inside the net. These insects need to be removed. If physical removal is not feasible, the standard insecticide space spray, 2-percent d-phenothrin (NSN 6840- 01-412-4634) can be used. The label and Material Safety Data Sheet (MSDS) for this product can be accessed at www.afpmb.org or by accessing the following link:

MSDS: http://www.afpmb.org/pubs/standardlists/msds/6840-01-412-4634_msds.pdf

Avoid breathing the pesticide vapors while spraying, and **DO NOT USE 2-PERCENT D-PHENOTHRIN ON THE SKIN OR CLOTHING**.

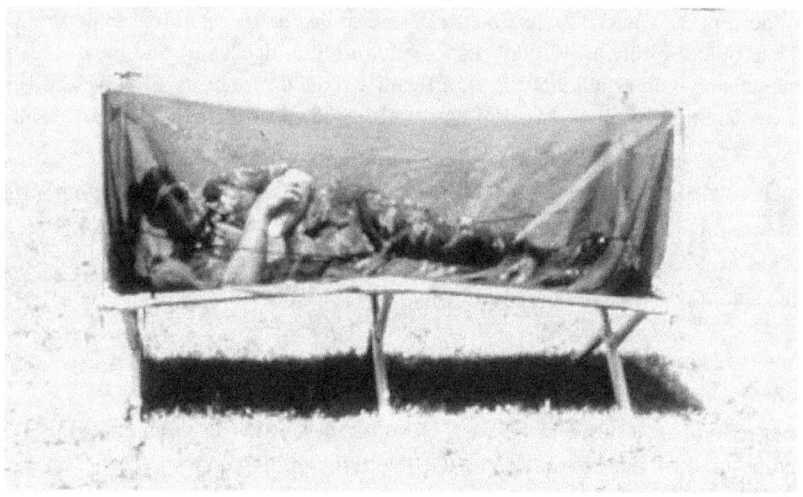

Figure 2-10. Insect net protector (Mosquito Bed Net), NSN 7210-00-266-9736 used with poles, (NSN 7210-00-267-5641), on folding cot.

(d) Before climbing inside, spray the net lightly with permethrin aerosol [see paragraph 2-9c(2)], or use a compressed air sprayer prior to setting it up [see paragraph 2-9c(3)(e) and Figure 2-24]. The permethrin will help protect against arthropods that are small enough to fit through the mesh of the net (e.g. sand flies). Allow the net to dry before handling. Permethrin applied by the 2-gallon sprayer (or compressed air sprayer) method to the bed net should provide protection for several months to a year or more (Personal communication, 1991; Loong et al., 1985).

(2) There are currently 3 types of pop-up bed nets on the NSN stock system:

- NSN 3740-01-516-4415: self-supporting low profile bed net (SSLPB); green camouflage. NO attachments for optional rain barrier; see photo and specifications at the following link:
    http://www.afpmb.org/pubs/standardlists/equipment/pdfs/3740-01-516-4415.pdf

- NSN 3740-01-518-7310: self-supporting low profile bed net (SSLPB); coyote brown with NO attachments for optional rain barrier; see photo and specifications at the following link:
    http://www.afpmb.org/pubs/standardlists/equipment/pdfs/3740-01-518-7310.pdf

- NSN 3740-01-543-5652: Improved bed net system; coyote brown. Equipped for addition of optional rain barrier; see photo and specifications at the following link: http://www.afpmb.org/pubs/standardlists/equipment/pdfs/3740-01-543-5652.pdf

These three pop-up bed nets, listed above, are identical in dimensions and all three are factory-treated with permethrin to provide long-lasting repellent protection against arthropods. They can either be used on top of a folding cot or set directly on the ground and used with or without a roll-up mat. Neither bed net NSN 3740-01-518-7310 or bed net NSN 3740-01-516-4415 can be combined with the optional rain shield (there are no attachment straps that connect to the rain shield) and therefore should only be procured for use when there is not any intent to order the rainshield components.

(a) NSN 3740-01-547-4426: this is the Enhanced Bed Net System (green camouflage). This combined unit comes with a pop-up bed net, rain shield, one Easton pole, 8 stakes, and repair kit. As a full-package item, it is intended to protect the occupant from most insects and also provide protection from rain and moisture. More information on this item, including photo, can be accessed at the following link: http://www.afpmb.org/pubs/standardlists/equipment/pdfs/3740-01-547-4426.pdf

(b) NSN 37440-01-546-4354: This is the Enhanced Bed Net System (Coyote Brown Camouflage). This combined unit comes with pop-up bed net NSN 3740-01-543-5652 (coyote brown with attachment straps), rain shield, one Easton pole, 8 stakes, and repair kit. As a full-package item, it is intended to protect the occupant from most insects and also provide proJ tection from rain and moisture. More information on this item, including photo, can be accessed at the following link: http://www.afpmb.org/pubs/standardlists/equipment/pdfs/3740-01-546-4354.pdf

Section IV. Repellents

2-7. Introduction

a. The concurrent use of repellents on the skin (deet) and clothing (permethrin) provides maximum personal protection against arthropods (McCain and Leach, 2007; Gupta et al., 1987; Lillie et al., 1988; Scholdt et al. 1988; Schreck and Kline, 1984; Schreck et al., 1989; Schreck et al., 1982). This dual strategy is known as the **DOD INSECT REPELLENT SYSTEM** and is explained in more depth at paragraph 2-11 below.

(1) Mosquitoes and certain other biting flies can bite either exposed skin or through light-weight clothing; whereas black flies, sand flies, biting midges, ticks, chiggers, and fleas may crawl underneath clothing to bite, in addition to biting exposed skin. Consequently, both types of treatments are necessary to provide maximum protection. (2) Clothing treatment with permethrin alone ordinarily does not adequately protect exposed skin because there is very

(2) Not all arthropod species are equally repelled by a particular repellent (Barnard and Xue, 2007). While deet is highly repellent to most mosquito and biting fly species, there are species of biting midges and mosquitoes, including certain species of *Anopheles* mosquitoes (malaria vectors), that are only partially repelled. Therefore, one should not discontinue using repellent if some bites are received when wearing deet, as other species that are present are still likely to be repelled. This example further illustrates the wisdom of utilizing the **DoD Insect Repellent System** [i.e., simultaneous use of both skin (deet) and clothing (permethrin) repellents].

(3) Some insect species are active during the day, others primarily at night. For this reason, it is important to follow recommendations provided by medical personnel, which may indicate the necessity of using repellents around the clock. Remember that lack of bites during the day, does not preclude the threat of attack during the evening or at night.

(4) Proper use of repellents may also reduce problems posed by filth flies and other nuisance pests. Unfortunately, no repellent appears to be significantly effective against stinging arthropods, such as bees, wasps, fire ants, and scorpions. The best strategy against them is simple avoidance.

2-8. Deet (N,N-diethyl-m-toluamide or N,N-diethyl-3-methylbenzamide).

a. Introduction

The standard skin repellent for the U.S. military since the mid 1950s has been the chemical N,N-diethyl-3-methylbenzamide that is commonly referred to as "deet" (Moore and Debboun, 2007; Frances, 2007). Deet is effective against a wide variety of arthropod species, especially mosquitoes and other biting flies, but also fleas, ticks, and chigger mites. In addition, deet has been reported to provide effective protection against land leeches, which are a problem primaril y in Southeast Asia.

b. Health and Safety Considerations

(1) Deet has been used safely for over 50 years by billions of people worldwide (Frances, 200 7). Although it has an excellent safety record, there have been isolated reports of harmful effects associated with its use. Most of these have been related to improper use, such as swallowing, spraying into the eye or applying to already irritated skin. While most of the complaints involve temporary minor skin or eye irritation, rare cases of toxic encephalopathy (inflammation of the brain) have been reported, but not confirmed, to be associated with deet usage, especially in young children. Other reported adverse reactions associated with, but not confirmed to be directly caused by deet, have included headache, nausea, behavioral changes, disorientation, loss of muscle coordination, irritability, confusion, and difficulty sleeping. While 200 million or more people use deet each year, there have been remarkably few reports of

(2) Since a small population of individuals may be sensitive to any chemical, it is important for personnel to apply repellents carefully following label instructions and to be aware of possible signs of intoxication. Apply deet evenly to exposed skin in accordance with label directions. Avoid contact with sensitive mucous membranes (e.g. eyes), the lips (accidental ingestion), and broken skin (e.g. abrasions, sunburn, poison ivy, existing insect bites).

(3) If the tactical situation permits, wash off deet repellent after the potential exposure to arthropods has ceased. Although deet is not soluble in water, it quickly washes off of skin, and out of clothing, with soap and water.

(4) Deet is a plasticizer and must be used with care to prevent damage to plastics, rubber, vinyl, or elastic items such as eyeglass frames, plastic lenses, and cases; contact lenses; combs; watch crystals; goggles; painted and varnished surfaces; and some synthetic fabrics (nylon excepted). The water-repellent properties of Gore-Tex® are also reduced by deet. Deet does **NOT** damage cotton or wool fabrics.

(5) More general information on deet can be obtained by accessing the following link:

http://deploymenthealthlibrary.fhp.osd.mil/products/DEET%20(245).pdf

c. <u>Formulations</u>

There are currently several deet-based products on the NSN list. The 3M products containing 33% deet, an extended-duration formulation is recommended as the standard military skin repellent.

(1) <u>Two-Ounce Tube</u>
(Insect Repellent, Personal Application, 3M Ultrathon ®, EPA Reg. No. 58007-1, extended-duration, NSN 6840-01-284-3982)(Figure 2-11)

(a) **THIS HAS BEEN THE MILITARY SKIN REPELLENT OF CHOICE** since 1990, when it first became available in the military supply system. It was developed by the Department of Defense in collaboration with the 3M Corporation. The product contains 33% deet in a controlled-release polymer base. It is applied as a non-greasy, white lotion with a mild, pleasant odor. The polymer in the formulation slows the absorption and evaporation of deet, thereby holding it on the surface of the skin where it can continue to repel arthropods for an 'extended' period of time. Laboratory testing shows that the extended-duration deet lotion provides 6 hours of at least 95-percent protection against a variety of mosquito species in a tropical environment, 10 hours in a hot, dry environment, and 12 hours in a forested/wet environment (Gupta and Rutledge, 1989)

(b) The label and Material Safety Data Sheet (MSDS) for this product can be accessed at the following link:

Label: http://www.afpmb.org/pubs/standardlists/labels/6840-01-284-3982_label.pdf
MSDS: http://www.afpmb.org/pubs/standardlists/msds/6840-01-284-3982_msds.pdf

Follow label directions. Dispense the lotion into one hand, rub the hands lightly together, and apply thoroughly in a thin layer over the forearms, upper arms, face, neck, ears, and other exposed areas. **DO NOT APPLY REPELLENT TO THE EYES AND LIPS, OR TO SENSITIVE OR DAMAGED SKIN** (for example, sunburn, abrasions, and poison ivy). Do not waste deet by applying it too thickly; a light, uniform coating provides excellent repellent protection.

Figure 2-11. Standard Military Skin Repellent, 33-Percent DEET, 2-Ounce Tube Extended-Duration, NSN 6840-01-284-3982

(c) If one begins to receive bites, reapply the repellent as described in paragraph (b), above. The value of the extended-duration formulation is that the polymer, by slowing loss of deet from the skin surface, retains deet at a concentration sufficient to repel arthropods for a long period of time [see paragraph (a), above]. Repellent formulations containing higher concentrations of deet do not provide longer, or better, repellency.

(d) The extended-duration deet formulation does **NOT** adversely affect the seal of the individual protective mask over the short term (USAMMDA Memorandum, 1989). However, the mask should be washed after each use to preclude damage to its surfaces by long-term exposure to residues of deet.

(e) The extended-duration deet formulation does **NOT** affect the infrared signature of the soldier (NRDEC Memorandum, 1991)

(f) The extended- duration deet formulation **CAN** be safely used with camouflage face paint; apply a thin layer of repellent from the tube and then follow with the face paint.

(g) Storage and disposal.

(1) This product is water-based and nonflammable. It is relatively heat and cold stable, although at high temperatures of over 140°F, some separation is possible and the product may begin to leak from the container. Under optimum conditions, shelf-life is five years or longer.

(2) After dispensing the contents, wrap the container in accordance with label instructions and discard in the trash. In contingency situations, follow appropriate operaJ tional guidance.

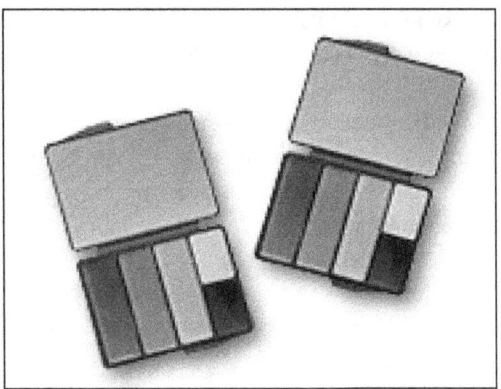

Figure 2-12.   NSN 6850-01-493-7309: Without deet (Green compact on left)
NSN 6840-01-493-7334: CONTAINS deet (Brown compact on right)

(a) This product contains 30% deet on a weight-by-weight basis to the other ingredients including the pigments and waxes that make up the camouflage paint. Therefore this formulation may not be as effective an arthropod repellent as the extended-duration deet lotion. The product is designed to provide camouflage face paint and repel mosquitoes, ticks, chiggers, deer flies, stable flies, black flies, fleas, and gnats. The product label and MSDS can be accessed at the following link:

Label: http://www.afpmb.org/pubs/standardlists/labels/6840-01-493-7334_label.pdf

MSDS: http://www.afpmb.org/pubs/standardlists/msds/6840-01-493-7334_msds.pdf

(b) Do NOT get in eyes or on lips. Do not apply over cuts, wounds or irritated skin.

(c) Storage and disposal: store in a cool, dry area away from heat. Do not reuse empty container; wrap container and put in trash or dispose of in accordance with operational guidance.

(3) Insect Repellent With Sunscreen, Personal Application, NSN 6840-01-288-2188, 2-oz tubes and NSN 6840-01-452-9582, packets (Figure 2-13)

Figure 2-13. Insect Repellent With Sunscreen, Personal Application, NSN 6840-01-288-2188, 2-oz tubes and NSN 6840-01-452-9582, packets

(a) These items are intended solely for use as a component of survival kits. The specific items stocked under these NSNs vary, based on the most suitable product available at the time. The product will contain both deet and sunscreen ingredients in various concentrations (e.g. 20-percent deet/Sun-Protection-Factor 15). The current label and MSDS for the 2-oz tube (NSN 6840-01-288-2188) can be accessed at the following link:

Label: http://www.afpmb.org/pubs/standardlists/labels/6840-01-288-2188_label.pdf
MSDS: http://www.afpmb.org/pubs/standardlists/msds/6840-01-288-2188_msds.pdf

The current label and MSDS for the packets (NSN 6840-01-452-9582) can be accessed at the following link:

Label: http://www.afpmb.org/pubs/standardlists/labels/6840-01-452-9582_label.pdf
MSDS: http://www.afpmb.org/pubs/standardlists/msds/6840-01-452-9582_msds.pdf

(b) Follow the label directions for use and disposal instructions.

(4) Insect Repellent Stick
(Personal Application, 30-percent deet, NSN 6840-00-142-8965)

TG No. 36   October 2009

(a) This product is a 1-ounce, waxy repellent stick marketed as Cutter® Insect Repellent Stick (Figure 2-14). It contains 30-percent deet in a waxy base. The label for this product can be accessed at the following link:

Label: http://www.afpmb.org/pubs/standardlists/labels/6840-00-142-8965_label.pdf
MSDS: http://www.afpmb.org/pubs/standardlists/msds/6840-00-142-8965_msds.pdf

It repels mosquitoes, biting midges, stable flies, sand flies, black flies, ticks, fleas, and chiggers. Its stable waxy formulation and convenient small size make it suitable for inclusion in survival kits.

Figure 2-14. Insect Repellent Stick, 30-Percent deet,

Personnel Application, 1-Ounce, NSN 6840-00-142-8965

(b) To use, push the stick up ½ inch. Apply over all exposed skin. Avoid the eyes and lips. For chiggers, fleas and ticks, also apply to the socks, tops of the shoes, and around all openings in the outer clothing. The product will not damage nylon, cotton, or wool. However, it may damage some synthetic fabrics, plastics, paints, and varnishes.

(c) Storage and disposal.

(1) The Cutter Insect Repellent stick is cold stable. At 140°F, however, the stick will begin to melt and leakage from the container can be expected (Personal communication, 1991). The product is not flammable, and under optimum conditions, the shelf-life is indefinite.

(3) After using the contents, wrap the container according to label instructions and discard in the trash. In contingency situations, dispose of in accordance with operational

2-9. Permethrin
[(3-phenoxyphenyl) methyl (+/-) cis/trans 3-(2,2-dichloroethenyl) 2,2-dimethyl-cyclopropanecarboxylate].

a. Introduction

(1) Permethrin is the U.S. military's standard repellent for application to fabric and is considered the most effective clothing impregnant available. McCain and Leach (2007) provided a concise history of clothing repellents in the U.S. military and fabric treatment by permethrin. The following link provides information on the health and safety aspects of permethrin impregnated clothing:

http://deploymenthealthlibrary.fhp.osd.mil/products/Permethrin%20-%20Impregnated%20Clothing%20(242).pdf

The primary mode of action is contact toxicity, particularly against crawling arthropods such as ticks (Evans et al., 1990; Mehr et al., 1986; Schreck et al., 1986; Mount and Snoddy, 1983; Schreck et al., 1982a+b), chigger mites (Breeden et al., 1982), fleas (Mehr et al., 1984), and lice (Nassif et al., 1980; Sholdt et al., 1989) Permethrin also acts as a contact repellent against mosquitoes and biting flies (Lindsay and McAndless, 1978; Schrek et al., 1978a; Schrek et al., 1978b; Sholdt et al., 1988), It is odorless, nonirritating, and resistant to reduction by washing and wear (Schrek et al., 1982; Schrek et al., 1980; Wirtz et al., 1985; Wirtz et al., 1986) Permethrin is bound so strongly to cotton or 50% cotton/nylon mix that repellency is still achieved even after 50 washings. After several washings, treated uniforms will continue to provide contact repellency, even though they may no longer be toxic to insects. **PERMETHRIN WILL NOT WASH OUT OF TREATED UNIFORMS WHEN WORN IN THE RAIN OR WHEN FORDING STREAMS, ETC.**

(2) Because it does not evaporate, permethrin does not provide protection to exposed skin adjacent to treated clothing. However, the concurrent use of repellents on the skin (DEET) and clothing (permethrin) should provide maximum personal protection against arthropods.

(3) It is important to realize there is science behind the testing and evaluation of permethrin impregnation of military fabrics and uniforms. Prospective textiles and uniforms are tested for binding efficiency with permethrin and then the subsequent bite protection that is conferred to the user. The following website link provides a summary of the evaluation processes that permethrin-impregnated uniforms undergo:

http://www.afpmb.org/pubs/tims/tg36/treateduniformevaluation.ppt

(4) Permethrin can be used to treat hot weather (100-percent cotton) and temperate (50-percent cotton/50-percent nylon; woodland or desert camouflage) military field uniforms. As of

field uniforms within the services. Not all uniforms can effectively be treated with permethrin! It is imperative that the user first research whether or not a subject uniform can be treated and what treatment method is approved. The following website address provides the current permethrin-treatment status of the various uniforms now in use:

http://www.afpmb.org/coweb/guidance_targets/ppms/Uniform%20Permethrin%20Treatment%2 0Matrixpdf

NOTE: On the document accessed by the above weblink, only those uniforms that are evaluated and approved for treatment by the specific applications are shown in the color green.

It is also recommended that the specific service's uniform authority be contacted for recommendations on permethrin guidelines. The following contact information for the respective uniform centers of each service is provided:

- Navy:
  Navy Uniform Matters Office
  N131U
  2 Navy Annex
  Washington, DC 20307
  (703) 614-5076/DSN 224
  http://WWW.NPC.NAVY.MIL/COMMANDSUPPORT/USNAVYUNIFORMS/

- Marine Corps:
  Permanent Marine Corps Uniform Board (PMCUB) 2200 Lester St
  Quantico VA 22134
  (703) 432-4607/4754 (DSN 378), FAX: (703) 432-3262
  http://www.marcorsyscom.usmc.mil/SITES/MCUB/

- Air Force:
  USAF Uniform Programs and Policy
  HQ USAF/A1SO
  201 12th Street South, Suite 413
  Arlington, VA 22202 AFA1SOU.
  Workflow@pentagon.af.mil
  DSN: 664-0147, commercial (703) 604-0147 (general questions)

- Army:
    Project Manager Soldier Equipment
    SFAE-SDR-SEQ
    10170 Beach Road Bldg 325
    Fort Belvoir, VA 22060-5862
    https://peosoldier.army.mil/
    AKO Website: https://www.us.army.mil/suite/page/389
    AKO Knowledge Center: https://www.us.army.mil/suite/community/190

(5) Once uniforms are treated, **DO NOT DRY-CLEAN PERMETHRIN-TREATED UNIFORMS**. The solvents used in the dry-cleaning process will remove the permethrin from the fabric as stated on the permethrin product labels.

(6) Other cloth items such as mosquito netting, camouflage helmet covers, ground covers, and tentage (with the exception of vinyl-coated temper tents) may also be treated in the field. Temper tents that have a vinyl-urethane finish cannot be treated with permethrin. Because the finish is water repellent, permethrin solutions will simply drip off.

b. Health and Safety Considerations

(1) The uniform cap should NOT be treated with permethrin because of the potential for excessive permethrin absorption through the scalp. Treatment of the cap is not critical since, due to its construction, it is considered impenetrable to biting insects.

(2) Do NOT treat underwear, including undershirts, or physical training uniforms. Although permethrin is poorly absorbed and is rapidly inactivated in mammals (McCain and Leach, 2007; Taplin and Meinking, 1990), wearing untreated undergarments signJ ificantly reduces the risk of exposure to fabric impregnants (USAEHA, 1982; USAEHA, 1 988a; USAEHA, 1988b).

(3) Precautionary measures should be observed when handling and mixing permethrin. Avoid permethrin contact with the face, eyes, and skin, and avoid breathing vapors or spray mist. Do not allow skin contact with treated surfaces until the chemical has dried completely. Wear protective gloves when handling wet, treated uniforms. In case of contact with the eyes, flush with plenty of water, and in case of contact with skin, wash with soap and water. Get medical attention if irritation persists. Do not allow the chemical to contact food, mess gear, or water supplies. Thoroughly wash dishes and utensils contaminated with permethrin.

(4) **THIS PESTICIDE IS EXTREMELY TOXIC TO FISH AND AQUATIC INVERTEBRATES**. Keep out of lakes, ponds, streams, or any waterways such as storm drains and irrigation ditches. Do not contaminate water by cleaning equipment or disposing of

c. Methods of permethrin treatment

Several different methods of permethrin treatment to fabric or material are available within the military supply system.   *(1) Individual Dynamic Absorption (IDA) Kit*
(Insect Repellent, Clothing Application, Permethrin, NSN 6840-01-345-0237)

(a) This item is a protective treatment kit for service field uniforms or equivalents that are approved for treatment with the IDA kit. This label and Material Safety Data Sheet (MSDS) can be accessed at the AFPMB website www.afpmb.org or at:

Label: http://www.afpmb.org/pubs/standardlists/labels/6840-01-345-0237_label.pdf
MSDS: http://www.afpmb.org/pubs/standardlists/msds/6840-01-345-0237_msds_hilton.pdf
or
http://www.afpmb.org/pubs/standardlists/msds/6840-01-345-0237_msds_sawyer.pdf

It provides excellent long-term protection (one treatment is effective for the life of the uniform). The IDA kit is sometimes referred to by the nicknames "baggie method" or "shake and bake." The kit contains materials sufficient to treat one complete uniform (shirt and trousers): two plastic vials of permethrin [40-percent emulsifiable concentrate (EC), 9-ml each], two plastic treatment bags, two pieces of twine, one pair of disposable protective gloves, and one black marking pen (one pen per four kits) (Figure 2-15).

(b) This item is perhaps the safest and most environmentally friendly field method by which individuals can treat their uniforms. An ideal way to train personnel on the correct use of this kit, is to provide instruction during a unit formation. The unit leadership can thus ensure the subject personnel have at least one treated uniform, and that each member knows how to treat additional uniforms. A big advantage of this method is that the IDA kits are compact and can be readily transported via air transport. The protection conferred by this permethrin treatment method is designed to last through 50 launderings of the garment.

(c) Wear the protective gloves when mixing to avoid accidental exposure to concentrated permethrin should spillage occur. Also wear protective eye wear such as safety glasses, eyeshield, or safety goggles; these items can be shared among users. Treat the uniform shirt and trousers separately, following the instructions printed on the back of each treatment bag (Figures 2-16 and 2-17).

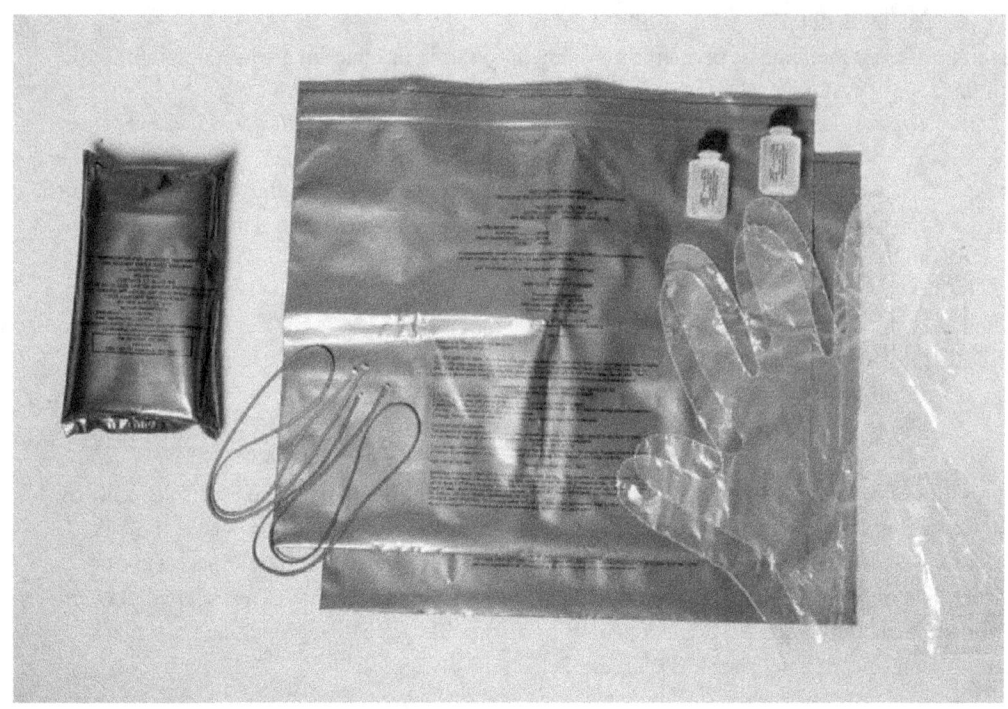

Figure 2-15. Permethrin IDA Kit, NSN 6840-01-345-0237 for treating a single field uniform with permethrin

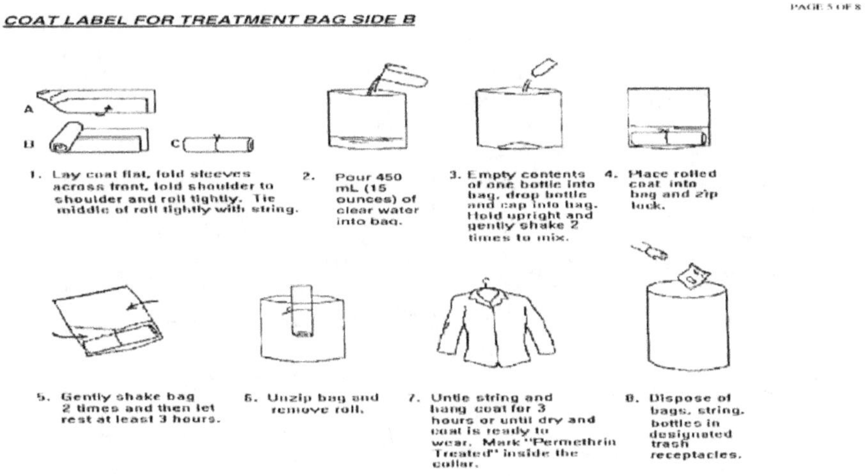

COAT LABEL FOR TREATMENT BAG SIDE B

1. Lay coat flat, fold sleeves across front, fold shoulder to shoulder and roll tightly. Tie middle of roll tightly with string.
2. Pour 450 mL (15 ounces) of clear water into bag.
3. Empty contents of one bottle into bag, drop bottle and cap into bag. Hold upright and gently shake 2 times to mix.
4. Place rolled coat into bag and zip lock.
5. Gently shake bag 2 times and then let rest at least 3 hours.
6. Unzip bag and remove roll.
7. Untie string and hang coat for 3 hours or until dry and coat is ready to wear. Mark "Permethrin Treated" inside the collar.
8. Dispose of bags, string, bottles in designated trash receptacles.

Figure 2-16. IDA kit instructions for treating coat half of the field uniform with permethrin, as they appear on Bag A of the IDA Kit.

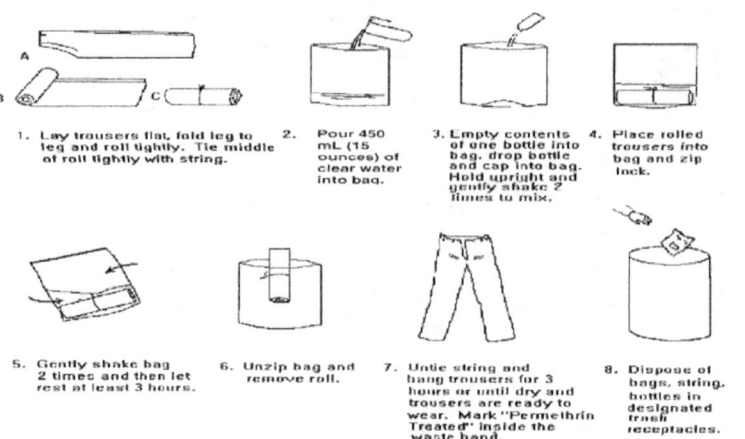

Figure 2-17. IDA kit instructions for treating trouser half of the field uniform with permethrin, as they appear on Bag B of the IDA Kit

(d) See Figures 2-18a-d for steps 1 through 4. Assemble all materials and don the proper personal protective gear (Step 1). Pour 450 milliliters (ml) of water (units will need to determine a reference for the 450 ml quantity and mark it on a readily available item such as a disposable plastic water bottle; when canteens were a common accouterment, 450 ml of water equaled ¾ of a canteen cup) into one of the bags (Step 2), add the contents of one of the vials of permethrin into the bag (Step 3), and gently agitate to mix (Step 4).

Figure 2-18a. Assemble material

Figure 2-18b. Don protective gear

Figure 2-18c. Add contents of one vial to water

Figure 2-18d. Agitate mixture

TG No. 36 October 2009

(e) Figures 2-19a-e detail the next steps. After rolling and tying the garment according to the instructions, place it in the bag (Step 5), re-seal the bag, agitate again (Step 6), and then allow the bag to be stationary for approximately 3 hours (Step 7). During this time, all the liquid is absorbed by the garment. Open the bag, remove the garment, and hang until dry which will take 2-4 hours, or more, depending upon weather conditions (Step 8).

(f) Do not reuse empty treatment bags. Place all used kit components into one treatment bag (Step 9), seal the bag, and put in the trash. In contingency situations, dispose of in accordance with operational guidance.

Figure 2-19a. Placing garment into bag

Figure 2-19b. Resealing bag

Figure 2-19c. Agitate with garmet

Figure 2-19d. Drying of garmet

Figure 2-19e. Place all used components into one treatment bag and then dispose

(g) As the uniform dries, the solvents, which are the source of the odor, evaporate. Once dry, permethrin has no odor and does not affect the appearance of the fabric. Even though the uniform may now be safely handled and worn, laundering the uniform one time is recommended. Permethrin is bound so strongly to the fabric by this procedure that water will not remove it: **PERMETHRIN WILL NOT WASH OUT OF TREATED UNIFORMS WHEN WORN IN THE RAIN OR WHEN FORDING STREAMS, ETC.** With the black pen, mark the inside coat collar and the inside waist band 'Perm treat, mo/yr' (this stands for "Permethrin treated, month/year").

(h) **DO NOT RE-TREAT THE UNIFORM:** one treatment is effective in preventing mosquito bites through the fabric for over 50 launderings. **DO NOT TREAT THE UNDERWEAR OR THE CAP.** Remember that dry-cleaning will completely remove permethrin.

(i) Starching field uniforms prior to treatment with permethrin does **NOT** adversely affect impregnation. Homogeneous absorption of permethrin is achieved in both hot and temperate-weather uniforms whether or not they are starched prior to treatment (NRDEC Memorandum, 1989).

(j) Permethrin-impregnated and untreated temperate-weather field uniforms **CAN** be laundered together. No significant transfer of permethrin from treated to untreated uniforms occurs during laundering (NRDEC Memorandum, 1989).

(k) Store the unused kits as described for the aerosol can in paragraph 2-9c(2)(c)(1-3) below. Under optimum conditions, the shelf life of this product is indefinite. If deterioration of the containers, and/or leakage of the contents, is detected prior to this time, turn in the

(l) This product is flammable and must be shipped in accordance with Department of Transportation (DOT) regulations.

(2) Aerosol Spray Can

(Insect Repellent, Clothing Application, Aerosol, Permethrin Arthropod Repellent, NSN 6840-01-278-1336)

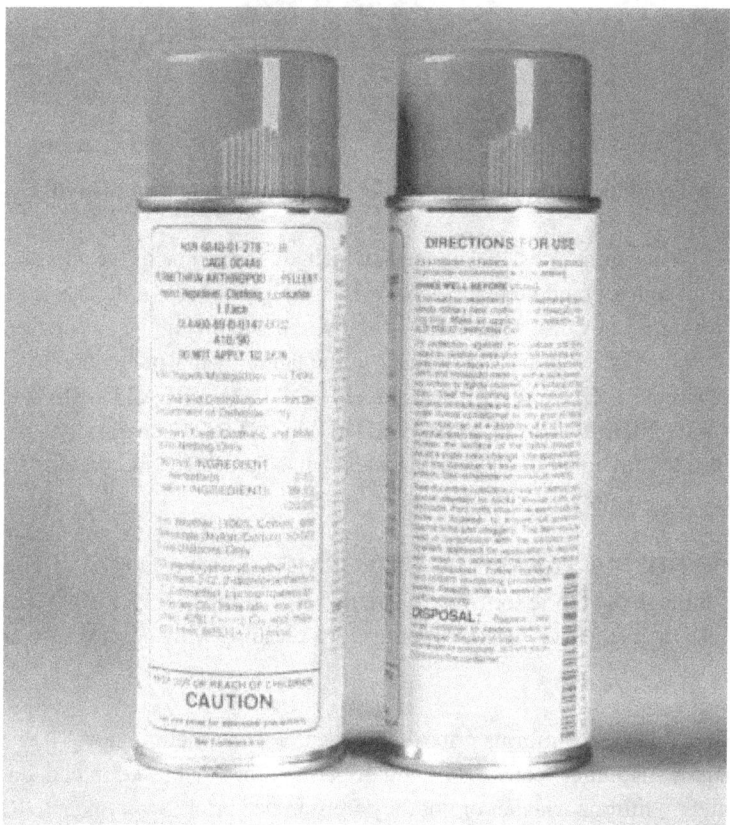

Figure 2-20. Permethrin aerosol, NSN 6840-01-278-1336, 6-ounce can
0.5 –percent Permethrin

(a) This product contains 0.5-percent permethrin in a 6-ounce can (Figure 2-20). The product label and MSDS can be accessed at the AFPMB website www.afpmb.org or clicking on this link:

Label: http://www.afpmb.org/pubs/standardlists/labels/6840-01-278-1336_label.pdf
MSDS: http://www.afpmb.org/pubs/standardlists/msds/6840-01-278-1336_msds.pdf

TG No. 36   October 2009

The aerosol spray can be used by the individual to treat field clothing (Figures 2-21a and b) as well as mosquito netting. Treatment of head nets (Figure 2-8) is considered mosquito netting and as such is acceptable. **DO NOT TREAT UNDERWEAR OR CAP**. This aerosol method of permethrin treatment is also available commercially under several different trade names.

(b) All applications should be made outdoors. Select a location protected from the wind. Shake well before using. Spray with a slow, sweeping motion while holding the can at a distance of 6 to 8 inches from the clothing (while not being worn).

(1) Spray the outer surfaces of the clothing, back and front, until the surface of the fabric appears moistened and a slight color change is noted (the original color will be restored when the uniform dries). Treat the shirt/blouse/coat and then the trousers, each for a minimum of 30 seconds on each side. Pay particular attention to the trouser cuffs and the shirt cuffs. Use approximately three-fourths of the can to treat one complete field uniform.

(2) The outer surface of the socks may also be **LIGHTLY** sprayed, regardless of whether they are cotton, wool, or a synthetic. The most critical areas are the top and front portions of the socks. This will aid in protecting against chiggers and tiny immature ticks which may find their way through the boot eyelets.

Figures 2-21a and 2-21b. Applying permethrin aerosol to the field uniform

(3) Allow the uniform to dry completely before being worn. This takes approximately 2 hours (or up to 4 hours under humid conditions). If possible, and if time permits, allow to dry in a shaded area because sunlight hastens degradation of permethrin. Permethrin has no odor once dry. Follow standard field uniform laundering procedures weekly. A drawback of this method is that reapplication is needed after 6 weeks or the sixth laundering, whichever comes first. Remember that dry-cleaning will completely remove permethrin.

TG No. 36   October 2009

(c) Storage and disposal.

(1) The aerosol should be stored at temperatures between 32°F and 130°F. At temperatures above 130oF there is increased chance of the can bursting. At temperatures below 32 °F, permethrin will begin to crystallize out of solution, although upon return to temperatures of 60-80°F, it re-dissolves with no apparent effect on the quality of the product (Personal communicaJ tion 1990). Under optimum storage conditions, the shelf-life of the aerosol is indefinite. Check permethrin containers for deterioration annually as evidenced by leakage or loss of propellant. Turn damaged product in for proper disposal.

(2) After the contents of the can have been dispensed, replace the cap, wrap the container in several layers of newspaper to provide a protective puncture buffer and discard in the trash per label instructions. Do not puncture or incinerate. In contingency situations, dispJ ose of in accordance with operational guidance.

(3) This product is **NOT** flammable, and may be safely carried aboard aircraft. Refer to DOT regulations for detailed guidance.

(3) 5.1-Ounce (151 ml) Bottle
(Insect Repellent, Clothing Application, Permethrin, 40-Percent Liquid, air compressed Sprayer
NSN 6840-01-334-2666).

(a) The product label and MSDS can be accessed at the AFPMB website www.afpmb.org or at the following link:

Label: http://www.afpmb.org/pubs/standardlists/labels/6840-01-334-2666_label.pdf
MSDS: http://www.afpmb.org/pubs/standardlists/msds/6840-01-334-2666_msds.pdf

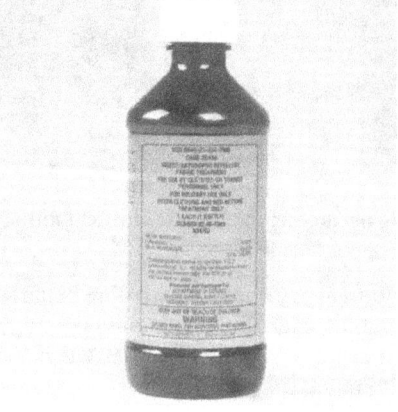

Figure 2-22. Permethrin 5.1-ounce (151-ml) bottle, NSN 6840-01-334-2666
insect repellent, clothing application, permethrin, 40-Percent Liquid, air-compressed sprayer

This product contains 40-percent permethrin EC (Figure 2-22). In accordance with its label, **IT IS FOR USE BY CERTIFIED OR TRAINED PERSONNEL ONLY**. This restriction must be carefully considered when determining what method of permethrin treatment is to be used. Are there sufficient personnel certified and up-to-date on appropriate occupational health testing such as respirator fit test to carry out this method of application? Is the equipment available?

(b) As per label instructions, this method can be applied to certain military field uniforms, netting, and tentage as approved by the military service. Treatment by this method will invariably result in runoff of the chemical. Wear appropriate personal protective equipment (e.g., respirator, gloves, etc.) when applying permethrin with this method. Ensure appropriate steps are taken to minimize/prevent contamination of the environment; water sources must not be contaminated.

(c) Application can be made by air compressed sprayers such as: 2-gallon sprayer (NSN 3740-00-641-4719), manually carried backpack sprayer (NSN 3740-01-496-9306), manually carried hydraulic backpack sprayer (NSN 3740-01-543-0676), manually carried compressed air backpack sprayer (NSN 3740-01-561-9663), and the battery powered sprayer (NSN 3740-01- 518-6876). For treatment of uniforms, the key requirement is that the sprayer must be calibrated to deliver 32 fluid ounces of mixture to each uniform (front and back). This delivery equals a fabric treatment level of 0.52% weight by weight of active ingredient.

(d) Several steps are essential in properly using this treatment method and product. Wear protective gloves and a respirator when mixing and applying this formulation. While a number of different air compressed sprayers can be used with this method (see above paragraph), the 2-gallon sprayer will be highlighted as the example:

(1) Thoroughly clean a 2-gallon sprayer by triple-rinsing with water. Add one gallon of clean water to the sprayer, followed by the entire contents of the 5.1 ounce bottle; then add a second gallon of water. This procedure helps to mix the water and permethrin. Agitate and bring to the maximum pressure required for the application as determined during the calibration process. (The 2-gallon sprayer, NSN 3740-00-641-4719, comes equipped with a pressure gauge. When retrofitting older sprayers, use NSN 3740-01-332-8746, gauge, and NSN 4330-01-332-1 639, filter/gauge.) As soon as spraying begins, if using a hand pumped, compressed air sprayer, the pressure will progressively drop, requiring frequent re- pressurizations.

(2) When using any of the approved air compressed sprayer, the sprayer must be calibrated to deliver a total of 32 fluid ounces of mixture to each uniform (front and back).

(e) To treat clothing (Figure 2-23), place the complete uniforms on the ground. One bottle of permethrin is enough to treat eight complete uniforms. Hang the uniforms until they are dry (usually 2-4 hours, during which time the original color will be restored). Once dry, the garments may be safely handled and worn. This procedure impregnates 8 sets of uniforms with permethrin. Permethrin is bound so strongly to the fabric by this procedure that water will not remove it: permethrin will not wash out of treated uniforms in the rain or when fording streams, etc.

Figure 2-23. Treating field uniforms with air-compressed sprayer

(f) **DO NOT RE-TREAT THE UNIFORMS**: one treatment is effective in preventing mosquito bites through the fabric for over 50 launderings. **DO NOT TREAT THE UNDERWEAR OR THE CAP**. Dry-cleaning will completely remove permethrin.

(g) To treat netting (Figure 2-24), spread the netting on the ground and spray on the netting for an even coverage. Spray with a slow sweeping motion to completely cover the netting fabric without runoff. Allow to dry completely before using. Re-treat after 1 year of use or six launderings. Bed nets that have been stored immediately following treatment will retain their effectiveness for many years.

Figure 2-24. Applying permetrhin by air-compressed sprayer to insect net protector (mosquito bed net)

       (h) To treat tentage that has not been coated with a water-repellent finish (Figure 2-25), erect the tent and treat the entryways and the inside surface (ceiling, walls and floor) as this is where pests are most likely to rest. Direct the spray to the walls, ceilings, and floor (if present) with a slow sweeping motion for even coverage just to the point of runoff. Permethrin is compatible with the fire retardants and mildew inhibitors used on general purpose, temper, and Arctic tents, as well as cotton tent liners. Re-treat after 9 months of use in temperate climates and after 6 months of use in tropical climates. Tents that have been stored following treatment will retain their effectiveness for many years. **PERMETHRIN SOLUTIONS ARE INEFFECTIVE ON VINYL-COATED TEMPER TENTS,** as the water-based permethrin will simply drip off of the water-repellent surface. In this case, it becomes even more important to use treated bed nets.

Figure 2-25. Applying permethrin by air-compressed sprayer to external surface of a tent Note: Permethrin solutions are ineffective on vinly-coated Temper Tents!

(i) Storage and disposal.

(1) Do not store products containing permethrin EC below 32°F, because the permethrin will crystallize. However, if that does occur, the integrity of the product can be restored when it is thawed, brought back to ambient temperature, and agitated until all the crystals redissolve. The flash point of 40-percent permethrin EC is 115°F due to the flammable solvent used in the formulation. Although the product shows little or no decomposition at 122°F after 30 days, storing the product in an enclosed space at or above 115°F will increase the chance of explosion due to ignition of vapors. Under optimum conditions, the shelf-life of this product is indefinite. However, if deterioration of the container occurs such as leakage, dispose of properly with guidance from an environmental/waste management authority.

(2) When empty, the pesticide container should be recapped, placed in a plastic bag, and discarded in the trash per label instructions. In contingency situations, dispose of in accordance with operational guidance.

(3) This product is flammable and must be shipped in accordance with DOT regulations.

(4) Factory Treatment of military uniforms

(a) In this method, military uniforms are factory-treated with permethrin prior to distribution. Factory-treated uniforms bear a unique label with a statement indicating that the fabric has been treated. Please access the following link for more information on factory permethrin treatment of military uniforms including the treatment of fire-retardant uniforms:

http://chppm-www.apgea.army.mil/documents/PermethrinFTUFS.pdf

(b) **DO NOT TREAT FACTORY-IMPREGNATED UNIFORMS WITH ADDITIONAL PERMETHRIN:** the original factory treatment is effective in preventing mosquito bites through the fabric for over 50 launderings. Remember that dry-cleaning will completely remove permethrin.

Summary on the permethrin treatment of uniforms
d.
The following steps should be taken when initiating the permethrin treatment of any military

STEP 1: Can the specific uniform be treated with permethrin?
Check the following link:
http://www.afpmb.org/coweb/guidance_targets/ppms/Uniform%20Permethrin%20Treatment%20Matrix.pdf

STEP 2: If it can be treated, what application method has been evaluated and approved?

In addition to the link provided in STEP 1, consult with the uniform authority for the respective service (contact information is listed under 2.9.a.(4))

STEP 3: Of the approved application methods, which one should be used?

Be careful when choosing a method as there are many factors to consider. For instance, perhaps one needs to treat a large quantity of uniforms and has a 2- or 3-month window prior to the deployment. The best option in this case may be to send the uniforms to a designated contractor for factory impregnation.

If the 5.1 ounce bottle method with air compressed sprayer is chosen, will there be a sufficient number of trained and fully certified personnel to conduct the spraying? Is there a proper application area available? Will there be sufficient drying time?

Another factor to consider is the expected time duration in the field. Is the deployment to last more than 6 weeks? If so, the spray can method may not be feasible as re-application is needed after 6 weeks or 6 launderings. If the spray operation has to take place in theatre, will there be available cargo for either the spray cans or 5.1 ounce bottles?

If IDA kits are used, will there be sufficient time for drying of the garments? The above are just some of the questions that must be asked when choosing a method; each situation will be different and must be decided on a case-by-case basis.

2-10. Miscellaneous Repellent.

CHIGG-AWAY® (Insect Repellent, Personal Application, NSN 6840-01-137-8456) is a yellow lotion with a sulfurous odor, available in a 188-ml plastic squeeze bottle (Figure 2-26). The label and Material Safety Data Sheet (MSDS) for this product can be accessed at www.afpmb.org or by clicking on this link:

Label: http://www.afpmb.org/pubs/standardlists/labels/6840-01-137-8456_label.pdf
MSDS: http://www.afpmb.org/pubs/standardlists/msds/6840-01-137-8456_msds.pdf

This product is labeled for use only against chiggers and as such is not recommended for use in most situations where other nuisance and/or vector arthropods may be active. The standard military repellents (extended-duration DEET lotion for skin and permethrin for clothing)

Figure 2-26. CHIGG-AWAY® 188-ml plastic squeeze bottle, NSN 6840-01-137-8456,
Insect Repellent
(Photo credited to: http://www.humco.com)

a. CHIGG-AWAY® contains 3-percent benzocaine to relieve itching caused by chigger and other insect bites, and 10% precipitated sulfur to repel chiggers. The product essentially combines a repellent (specific only against chiggers) and a soothing topical for relief from insect bites. To relieve itching, it can be applied directly to the bites of chiggers, mosquitoes, ticks, sand fleas and biting flies, or to skin irritation caused by poison ivy/oak and sunburn. As a repellent, it should be applied around feet, ankles, waist, to skin under all areas of light clothing, and around all openings in outer clothing. This product washes off easily, so reapply after heavy perspiration.

b. Do not apply this product to the eyes or other mucous membranes. The product is NOT intended for prolonged use.

c. Storage and Disposal.

(1) The shelf life of CHIGG-AWAY® is approximately 4 years. An expiration date is stamped on the container. It should be stored at room temperature, not above 100°F and should be kept from freezing. It is non-flammable and non-reactive.

(2) The empty bottle should be rinsed with tap water and discarded in the trash. In contingency situations, the container can be disposed of in the same manner as other

TG No. 36 October 2009

2-11. DoD Insect Repellent System

The **BEST STRATEGY** for defense against insects and other diseases-bearing arthropods is the **DOD INSECT REPELLENT SYSTEM** that is summarized in the following link:

http://chppm-www.apgea.army.mil/documents/FACT/DODInsectRepellentSystemJustheFacts-June2007.pdf

This system includes the application of extended-duration deet lotion to exposed skin, coupled with the application of permethrin to the field uniform. When used with a properly-worn uniform, the DoD insect repellent system will provide nearly complete protection from arthropod-borne diseases. This system should always be augmented with tick/insect checks on clothing and exposed skin.

2-12. Area Repellents.

    a. Area repellents include products that prevent bites over a large area rather than just on a person or their clothing. Some products claim to prevent bites by emitting sounds or electromagnetic waves. Other products use various methods of dispersing chemical compounds into the air.

    (1) Products include candles, burning coils, heat dispersed chemicals (from electric elements, butane combustion, or a candle), and vermiculite impregnated with various chemicals. Efficacy under ideal conditions varies from nearly complete prevention of bites to no protection at all. For devices that emit chemical compounds, protective effects are greatly affected by wind, and product claims for a given area of protection are based on conditions without any breezes.

    (2) Protection areas can decrease to virtually zero on the upwind side of these area repellent devices. Thus, area repellents may give a false sense of security to persons in the vicinity resulting in their not using skin and clothing repellents.

    b. The AFPMB does NOT recommend the use of any area repellent as an effective means for repelling arthropods, especially in a field environment. Currently, there are no area repellents listed on the NSN list. It is recommended that only the pest management items listed on the NSN list (can be accessed at link: http://www.afpmb.org/standardlist.htm) be procured for use in any and all military operations.

2.13 Repellent Devices that are Worn on the Body

    a. Wrist bands, broaches, etc. that contain repellent compounds and that are supposed to prevent insects from biting the wearer, are also of little use. Wrist bands have their effect only in the immediate vicinity of the band itself (i.e., the wrist or forearm of a person wearing one).

b. Broaches and pins containing repellents are likewise as limited. Additionally, simply walking around creates sufficient breeze across a person's body to reduce the effectiveness of these devices.

c. Soldiers, Airmen, Sailors, and Marines have used animal flea and tick collars, and cattle ear tags around their wrists, ankles, arms, or tied to belt lines or boots or boot laces in the past. Such products contain many different kinds of pesticides which may have adverse dermal and/or systemic effects on people. Severe skin reactions have been reported from using these products (Figure 2-27). In addition, some pesticides contained in these collars could trigger chemical agent detectors.

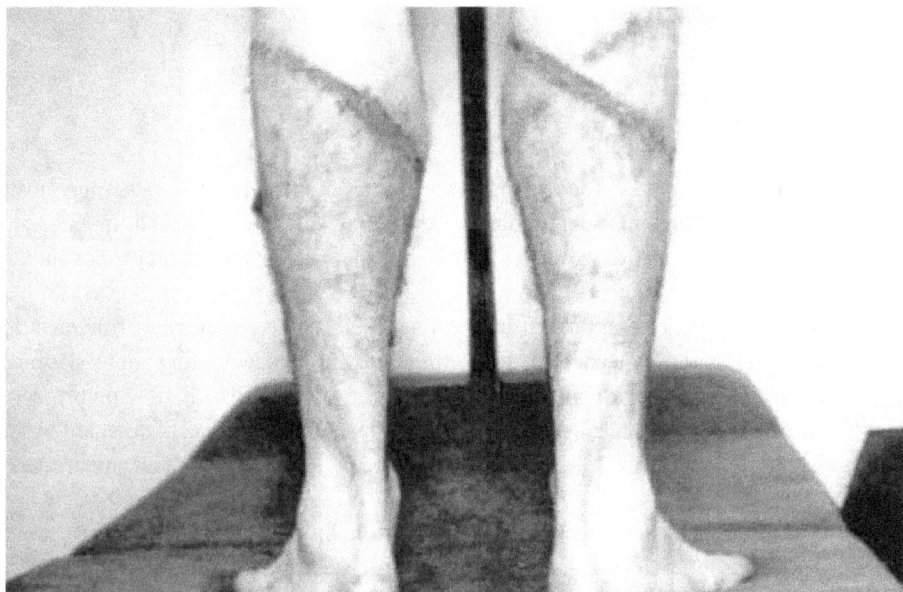

Figure 2-27. Skin lesions on the legs caused by human use of flea and tick collars

These collars and ear tags are **NOT INTENDED FOR HUMAN USE** so their safety has never been tested on humans.

TG No. 36   October 2009

Section V. Ineffective and Hazardous Practices

2-14. Introduction

a. A number of commercial products which are either not marketed for personal protection, or are not very effective repellents, are nevertheless being widely used by troops for this purpose. Such products are less effective than those containing deet, and they may be hazardous when used in a manner not approved by the label. Products with concentrations in the range of about 20% to 40% deet provide an appropriate mix of effectiveness and duration of protection. Within this range of concentration, percent active ingredient generally translates to duration of protection.

b. Products with less than 20% deet provide protection for too short a period of time to be useful in military scenarios. Above 40% deet, the gain in duration of protection is generally not great enough to warrant using the higher concentration products. Medical personnel should instruct troops on the correct use of appropriate personal protective measures and should strictly prohibit the use of unauthorized products.

2-15. Commercial Products

a. Non-deet Products

Many new repellent products that do not contain deet have recently become available. Some may use active ingredients that are not proven repellents or may have very little if any repellency (e.g., bath oils and some "natural" products); others may contain effective repellent ingredients at low concentrations such that they are effective for relatively short periods of time and therefore are not useful under field conditions. Some products, NOT intended for or registered for human use may actually be hazardous to use on humans (e.g., flea/tick collars and cattle ear tags). While non-deet products registered with the U.S. EPA and recommended by the U.S. CDC will provide reasonable protection for limited duration exposure within the U.S. or for tourists traveling internationally, their repellency is not equal to the NSN stocked deet repellents to protect personnel in the field against arthropod vectors that may carry disease.

b. Ingested Products

Some products or publications make claims that ingesting certain materials will protect you from insect bites. There is no scientific evidence tat any material that is ingested (e.g., match heads, vitamin B1, etc.) has any repellent effect on insects or other biting arthropods.

c. Recommendation

It is again stressed that only products listed on the DoD Standard Pesticides and Pest Control Equipment Lists (http://www.afpmb.org/standardlist.htm) should be used for any personal proJ tection item as these items have been through extensive and exhaustive evaluations and found to

TG No. 36   October 2009

Section VI. Pesticide Reduction through Physical/Mechanical Modifications and Sanitation

2-16. DoD and IPM

a. Current DoD environmental policy stresses a concerted effort to reduce pesticide use within military programs. As a part of this overall effort, use of the DOD INSECT REPELLENT SYSTEM will help to reduce the need for pesticide applications during contingency operations, as well as during routine training activities. The link between repellent use and reduced need for pesticide dispersal is nothing new, but has become increasingly more important in this era of enlightened environmental stewardship. Pesticide treatment may be necessary when troops are to remain for a prolonged period of time in an area that is heavily infested with arthropods. Pesticide applications must only be performed by trained or certified individuals, and only after Preventive Medicine personnel determine that other protective and preventive measures are, or will not be, fully successful. Aerial applications may be an option for extensive areas, but will need to be conducted by certified personnel in accordance to all applicable environmental laws and regulations.

b. The DoD is committed to the concept of Integrated Pest Management, an effective and environmentally sensitive approach to pest management that relies on a combination of common- sense practices. Physical/mechanical modifications such as clearing away leaf litter and underbrush that provide potential habitat for arthropods and/or for animal hosts may reduce nuisance and vector arthropod numbers. Raking is simple and efficient. Keep grass and weeds mowed back where possible, especially around buildings, and in housing, cantonment and recreational areas.

c. For large areas, controlled burning of the under story may be necessary although this should only be attempted by trained personnel when other methods fail or are impractical, and after authorization has been obtained through appropriate environmental and medical channels.

d. Mosquito breeding sites should be eliminated or reduced by draining standing water, and by preventing water accumulation in containers, depressions in the ground, or other recJ eptacles. Although not primarily a personal protective measure, it is the responsibility of each individual to participate in the overall unit sanitation effort. Once a bivouac site is established, sanitation is important.

e. Garbage and other odiferous decaying matter will attract arthropods and other animal pests and should not be allowed to accumulate. These types of materials should be maintained in tightly closed containers, or should be buried, burned, or removed.

Section VII. Conclusion

2-17. Summary

Conscientious use of the DOD INSECT REPELLENT SYSTEM, and the other protective

2-18. Training Package

A Power Point file providing a training package that covers the salient points of this TG can be accessed at the following link:

http://chppm-www.apgea.army.mil/documents/PersonalProtectiveMeasuresApril2005.ppt

or by linking into the general USACHPPM website and then selecting the most recent posted presentation for personal protection training:

http://chppm-www.apgea.army.mil/DoDIRS/

TG No. 36

APPENDIX A - REFERENCES

Barnard, D.R. and R. Xue. Biometrics and Behavior in Mosquito Repellent Assays. In: Insect Repellents: Priniciples, Methods, and Uses. Editors: Debboun, M., Frances, S. and Strickman, D. 2007. CRC Press. ISBN 0-8493-7196-1

Breeden, G.C., C.E. Schreck and A.L. Sorensen. 1982. Permethrin as a clothing treatment for personal protection against chigger mites (Acarina: Trombiculidae). Am. J. Trop. Med. Hyg. 31(3):589-592.

Bunn, R.W., K.L. Knight, and W. J. LaCasse. 1955. The Role of Entomology in the Preventive Medicine Program of the Armed Forces. Military Medicine. 116(2):119-124.

Dickens, T. September/October 1990. Vector Control as a Force Multiplier. Defense 90. pp. 26-35.

Elridge, B.F. and Edman, J.D. 2004. Medical Entomology: A Textbook on Public Health and Veterinary Problems Caused by Arthropods. Kluwer Academic Publishers.

Evans, S.R., G.W. Korch, Jr and M.A. Lawson. 1990. Comparative field evaluation of permethrin and DEET-treated military uniforms for personal protection against ticks (Acari). J. Med. Entomol. 27(5):829-834.

Frances, S.P. 2007. Efficacy and safety of repellents containing Deet. In: Insect Repellents: Priniciples, Methods, and Uses. Editors: Debboun, M., Frances, S. and Strickman, D. 2007. CRC Press. ISBN 0-8493-7196-1

Gambel, J. 1995. Preventing Insect Bites in the Field: A Key Force Multiplier. Army Med. Dept. Journal. PB 8-95-5/6 May-June. pp. 34-40.

Goddard, J. 2007. 5th ed. Physician's guide to arthropods of medical importance. CRC Press, Boca Raton, FL. 457 p.

Gupta, R.K., A.W. Sweeney, L.C. Rutledge, R.D. Cooper, S.P. Frances and D.R. Westrom. 1987. Effectiveness of controlled-release personal-use arthropod repellents and permethrin-impregnated clothing in the field. J. Am. Mosq. Control Assoc. 3(4):556-560.

Gupta, R.K. and L.C. Rutledge. 1989. Laboratory evaluation of controlled-release repellent formulations on human volunteers under three climatic regimens. J. Am. Mosq. Control Assoc.

Heymann, D. (ed.) 2004. 18th ed. Control of Communicable Diseases Manual. Amer. Public Health Assn., Washington, DC. 701 pp.]

Jinjiang, X., Z. Meiluan, L. Xinfu, G. Rongen, P. Shixian and L. Shuyou. 1988. Evaluation of permethrin-impregnated mosquito-nets against mosquitoes in China. Med. and Vet. Entomol. 2 :247-251.

Kettle, D. S. 1995. Medical and Veterinary Entomology. 2nd edition. CAB International.

Korzeniewski, K. and R. Olszanski. 2004. Leishmaniasis among soldiers of stabilization forces in Iraq. Int. Marit. Health. 55(1-4): 155-63.

Lillie, T.H., C.E. Schreck and A.J. Rahe. 1988. Effectiveness of personal protection against mosquitoes in Alaska. J. Med. Entomol. 25(6):475-478.

Lindsay, I.S. and J.M. McAndless. 1978. Permethrin-treated jackets versus repellent-treated jackets and hoods for personal protection against black flies and mosquitoes. Mosq. News. 3 8(3):350-356.

Loong, K.P., S. Naidu, E.S. Thevasagayam, and W.H. Cheong. 1985. Evaluation of the effectiveness of permethrin and DDT-imgregnated bed-nets against _Anopheles maculatus_. Southeast Asian J. Trop. Med. Publ. Hlth. 16(4):554-559.

McCain, W.C. and G.L. Leach. 2007. Repellents Used in Fabric: The Experience of the U.S. Military. In: Insect Repellents: Priniciples, Methods, and Uses. Editors: Debboun, M., Frances, S. and Strickman, D. 2007. CRC Press. ISBN 0-8493-7196-1

Mehr, Z.A., L.C. Rutledge and J.L. Inase. 1984. Evaluation of commercial and experimental repellents against Xenopsylla cheopis (Siphonaptera: Pulicidae). J. Med. Entomol. 21(6):665-669.

Mehr, Z.A., L.C. Rutledge, E.L. Morales and J.L. Inase. 1986. Laboratory evaluation of commercial and experimental repellents against Ornithodoros parkeri (Acari: Argasidae). J. Med. Entomol. 23(2):136-140.

NRDEC (Natick Research, Development and Engineering Center) Memorandum STRNC-IUC, 8 April 1991, subject: Effects of the New Insect Repellent, Personal Application (Cream Lotion, Tube); MIL-I-44415, on Battle Dress Uniform Material

NRDEC (Natick Research, Development and Engineering Center) Memorandum STRNC-ITCP, 2 8 December 1989, subject: Interim Progress Report on Arthropod Repellent Impregnant

Mount, G.A. and E.L. Snoddy. 1983. Pressurized sprays of permethrin and DEET on clothing for personal protection against the Lone Star tick and the American dog tick (Acari: Ixodidae). J. Econ. Entomol. 76:529-531.

Moore, S.J. and M. Debboun. 2007. History of Insect Repellents. In: Insect Repellents: Principles, Methods, and Uses. Editors: Debboun, M., Frances, S. and Strickman, D. 2007. CRC Press. ISBN 0-8493-7196-1

Mullen, G. and L. Durden, Editors. 2009. Medical and Veterinary Entomology. 2nd edition. Academic Press. ISBN: 978-0-12-372500-4.

Nassif, M., J.P. Brooke, D.B.A. Hutchinson, O.M. Kamel and E.A. Savage. 1980. Studies with permethrin against bodylice in Egypt. Pestic. Sci. 11:679-684.

Personal communication between Sandra Evans, USAEHA, and Carl Schreck, USDA, 11 March 1991, subject: Permethrin re-treatment regimens for military uniforms, mosquito metting, and tents.

Personal communication between Sandra Evans, USAEHA, and Brian Spesard, Miles, Inc., 15 January 1991, subject: Stability of Cutter Insect Repellent Stick.

Personal communication between Sandra Evans, USAEHA, and Paul Schoenberg, Fairfield American Corporation, 16 August 1990, subject: Flammability and shelf-life of permethrin 40-percent EC.

Plorde, J.J. 1983. Chapter 233, Scabies, Chiggers, and Other Ectoparasites. Harrison's Principles of Internal Medicine. 10th ed. McGraw-Hill, Inc. pp. 1239-1241.

Rossignol, P.A. and R.M. Feinsod. 1990. Chapter 58, Arthropods Directly Causing Human Injury. Tropical and Geographical Medicine. 2nd ed. McGraw-Hill, Inc. pp. 519-532.

Schreck, C.E. 1989a. Chapter 21, Protection from Blood-Feeding Arthropods, Management of Wilderness and Environmental Emergencies, ed 2, St. Louis, C.V.Mosley Company, pp 589-602.

Schreck, C.E. and D.L. Kline. 1989b. Personal protection afforded by controlled-release topical repellents and permethrin-treated clothing against natural populations of Aedes taeniorhynchus. J. Am. Mosq. Control Assoc. 5(1):77-80.

Schreck, C.E. and T.P. McGovern. 1989c. Repellents and other personal protection strategies against Aedes albopictus. J. Am. Mosq. Control Assoc. 5(2):247-250.

extended duration controlled release repellent formulations as personal protection against biting arthropods of military importance.

Schreck, C.E., E.L. Snoddy and A. Spielman. 1986b. Pressurized sprays of permethrin or DEET on military clothing for personal protection against Ixodes dammini (Acari: Ixodidae). J. Med. Entomol. 23(4):396-399.

Schreck, C.E., D.G. Haile and D.L. Kline. 1984. The effectiveness of permethrin and DEET, alone or in combination, for protection against Aedes taeniorhynchus. Am. J. Trop. Med. Hyg. 33(4):725-730.

Schreck, C.E., D.L. Kline, B.N. Chaniotis, R.N. Wilkinson, T.P. McGovern and D.E. Weidhaas. 1982a. Evaluation of personal protection methods against Phlebotomine sand flies including vectors of leishmaniasis in Panama. Am. J. Trop. Med. Hyg. 31(5):1046-1053.

Schreck, C.E., G.A. Mount and D.A. Carlson. 1982b. Wear and wash persistence of permethrin used as a clothing treatment for personal protection against the Lone Star tick (Acari: Ixodidae). J. Med. Entomol. 19(2):143-146.

Schreck, C.E., G.A. Mount and D.A. Carlson. 1982c. Pressurized sprays of permethrin on clothing for personal protection against the Lone Star tick (Acari: Ixodidae). J. Econ. Entomol. 75(6):1059-1061.

Schreck, C.E., D.A. Carlson, D.E. Weidhaas, K. Posey and D. Smith. 1980a. Wear and aging tests with permethrin-treated cotton-polyester fabric. J. Econ. Entomol. 73(3):451-453.

Schreck, C.E., E.L. Snoddy and G.A. Mount. 1980b. Permethrin and repellents as clothing impregnants for protection from the Lone Star tick. J. Econ. Entomol. 73(3):436-439.

Schreck, C.E., K. Posey and D. Smith. 1978a. Durability of permethrin as a potential clothing treatment to protect against blood-feeding arthropods. J. Econ. Entomol. 71(3):397-400.

Schreck, C.E., N. Smith, D. Weidhaas, K. Posey and D. Smith. 1978b. Repellents vs. toxicants as clothing treatments for protection from mosquitoes and other biting flies. J. Econ. Entomol. 71(6):919-922

Sholdt, L.L., E.J. Rogers, Jr, E.J. Gerberg, and C.E. Schreck. 1989. Effectiveness of permethrin-treated military uniform fabric against human body lice. Mil. Med. 154(2):90-93.

Spielman, A. and A.A. James. 1990. Chapter 20, Transmission of Vector-Borne Disease, Tropical and Geographical Medicine, 2nd Ed. McGraw-Hill, Inc. pp. 146-159.

Taplin, D. and T.L. Meinking. 1990. Pyrethrins and pyrethroids in dermatology. Arch. Dermatol. 126:213-221.

USACHPPM (U.S. Army Center for Health Promotion and Preventive Medicine). 2004. Insect Repellent System Poster

USAEHA (U.S. Army Environmental Hygiene Agency), Memorandum HSHB-MO-T, 13 September 1988 (a), subject: Phase 2, Migration of Permethrin from Military Fabrics Under Varying Environmental Conditions, Study No. 75-52-0687-88.

USAEHA (U.S. Army Environmental Hygiene Agency), Memorandum HSHB-MO-T, 12 April 1988 (b), subject: Fabric/Skin Contact from Wearing the Army Battle Dress Uniform, Study No. 75-52-0687-88, June - July 1987.

USAEHA (U.S. Army Environmental Hygiene Agency), Letter HSHB-LT-P/WP, 16 July 1982, subject: Interim Report, Migration of ^{14}C Permethrin from Impregnated Military Fabric, Study No. 75-51-0351-82, December 1981 - February 1982.

USAMMDA (U.S. Army Medical Materiel Development Activity) Memorandum, , SGRD-UMB, 4 April 1989, subject: Protocol Entitled "Effect of Extended Duration Topical Insect/ Arthropod Repellent on Fit of the Individual Protective Mask, M17," Submitted by LTC Lyman W. Roberts, MS, USAMMDA (Log No. A4893).

Wirtz, R.A., L.W. Roberts, J.A. Hallam, L.M. Macken, D.R. Roberts, M.D. Buescher and L.C. Rutledge. 1985. Laboratory testing of repellents against the tsetse Glossina morsitans (Diptera: Glossinidae). J. Med. Entomol. 22(3):271-275.

Wirtz, R.A., E.D. Rowton, J.A. Hallam, P.V. Perkins and L.C. Rutledge. 1986. Laboratory testing of repellents against the sand fly Phlebotomus papatasi (Diptera: Psychodidae). J. Med. Entomol. 23(1):64-67.

World Health Organization (WHO). 1989. Self-protection and vector control with

www.ingramcontent.com/pod-product-compliance
Lightning Source LLC
Chambersburg PA
CBHW080824170526
45158CB00009B/2515